Contents

Words appearing in the text in bold, like this, are explained in the Glossary.

What are carbon, oxygen, and nitrogen?

Carbon, oxygen, and nitrogen are all **elements**. We all need carbon, oxygen, and nitrogen in order to stay alive. You may know that we need oxygen to breathe, but are carbon and nitrogen just as important? The answer is yes, as without them there would be no food for you to eat!

Carbon is a solid, while oxygen and nitrogen are gases. When carbon and oxygen are mixed together they combine to form carbon dioxide, which is also a gas. Carbon dioxide, oxygen, and nitrogen are all found in the **atmosphere**. This is the blanket of air that surrounds the Earth. All three gases are necessary for life on Earth.

The air around us contains carbon dioxide, oxygen, and nitrogen.

Carbon-Oxygen and Nitrogen Cycles

Rebecca Harman

Heinemann

www.heinemann.co.uk/library
Visit our website to find out more information about Heinemann Library books.

To order:
☎ Phone 44 (0) 1865 888066
🖹 Send a fax to 44 (0) 1865 314091
🖥 Visit the Heinemann Bookshop at www.heinemann.co.uk/library to browse our
catalogue and order online.

First published in Great Britain by Heinemann
Library, Halley Court, Jordan Hill, Oxford
OX2 8EJ, part of Harcourt Education.
Heinemann is a registered trademark
of Harcourt Education Ltd.

Editorial: Melanie Copland
Design: Victoria Bevan and AMR Design
Illustration: David Woodroffe
Picture Research: Mica Brancic and
Helen Reilly
Production: Victoria Fitzgerald

Originated by Chroma Graphics
(Overseas) Pte. Ltd
Printed in China by WKT Copmany Limited

The paper used to print this book comes
from sustainable resources.

ISBN 13: 978 0431 013039 (HB)
ISBN 10: 0 431 01303 9 (HB)
09 08 07 06 05
10 9 8 7 6 5 4 3 2 1

ISBN 13: 978 0431 013121 (PB)
ISBN 10: 0 431 01312 8 (PB)
09 08 07 06
10 9 8 7 6 5 4 3 2 1

Acknowledgements
The Publishers would like to thank the following
for permission to reproduce photographs:
Alamy Images/Jason Friend **p.9**; Alamy
Images/Agripictures **p.6**; Alamy Images/Ron
Scott **p.16**; Alamy Images/Oote Boe **p.19**;
Alamy Images/Ashley Cooper **p.25**; Corbis
pp.20, 26: Corbis/Gary Braasch **p.8**; Corbis/Tim
Davies **p.13**; Corbis/Larry Lee Photography
p.21; Digital Vision **pp.4, 12**; Getty
Images/PhotoDisc **pp.17, 18, 22, 29**;
Science Photo Library **p.27**; Still Pictures/SJ
Krasemann **p.10**.

Cover photograph of tree stump in British
Columbia, Canada reproduced with
permission of Corbis/Gunter Marx.

The Publishers would like to thank Nick
Lapthorn for his assistance in the preparation
of this book.

Disclaimer
All Internet addresses (URLs) given in this
book were valid at the time of going to press.
However, due to the dynamic nature of the
Internet, some addresses may have changed,
or sites may have changed or ceased to exist
since publication. While the author and
Publishers regret any inconvenience this may
cause readers, no responsibility for any such
changes can be accepted by either the
author or the Publishers.

Every effort has been made to contact
copyright holders of any material reproduced
in this book. Any omissions will be rectified in
subsequent printings if notice is given to the
Publishers.

British Library Cataloguing in Publication Data
Harman, Rebecca
Carbon-Oxygen and Nitrogen Cycles:
respiration, photosynthesis and decomposition
(Earth's Processes)
577.1'4

A full catalogue record for this book is
available from the British Library.

There is only a tiny amount of carbon dioxide in the atmosphere. Although it makes up only about 0.03 per cent of the atmosphere, it is a very important gas. It is needed for plants to grow and to produce fruit and vegetables, which animals and humans can then eat.

Oxygen makes up about 21 per cent of the atmosphere. It is necessary for life. All plants and animals need oxygen to break down food and release the energy needed for movement and growth.

Nitrogen makes up a massive 78 per cent of the atmosphere. It is needed by all living things to make **proteins**, which provide energy for growth.

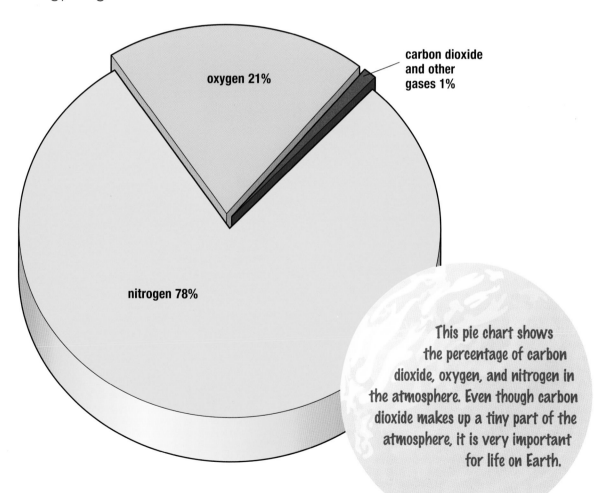

oxygen 21%

carbon dioxide and other gases 1%

nitrogen 78%

This pie chart shows the percentage of carbon dioxide, oxygen, and nitrogen in the atmosphere. Even though carbon dioxide makes up a tiny part of the atmosphere, it is very important for life on Earth.

Carbon dioxide also has an important effect on **climate**. It **absorbs** heat in the atmosphere, helping to keep the Earth warm. This makes all life on Earth possible. Without carbon dioxide in the atmosphere, the Earth would be too cold to live on.

Carbon, oxygen, and nitrogen are being changed into different forms all the time. They are **recycled** between the soil, plants, animals, the air, and the ocean. Without this continuous recycling, life on Earth would gradually grind to a halt.

Carbon, oxygen, and nitrogen are recycled between the grass, the cow, and the atmosphere.

What is the carbon-oxygen cycle?

The movement of carbon and oxygen between the atmosphere, the oceans, plants, animals, and the ground is called the **carbon-oxygen cycle**. This cycle is very important as it makes sure that the atmosphere always contains the right amount of carbon dioxide and oxygen.

In the carbon-oxygen cycle there are a number of **sinks**. These are places where carbon or oxygen are removed from the cycle and stored. Carbon and oxygen may be stored for millions of years, such as when they become locked up in **limestone** rocks. They may be stored for just a few days, such as when plants absorb carbon dioxide and return some of it, together with oxygen, to the air.

Carbon and oxygen are always moving between the atmosphere, the oceans, living things, and rocks.

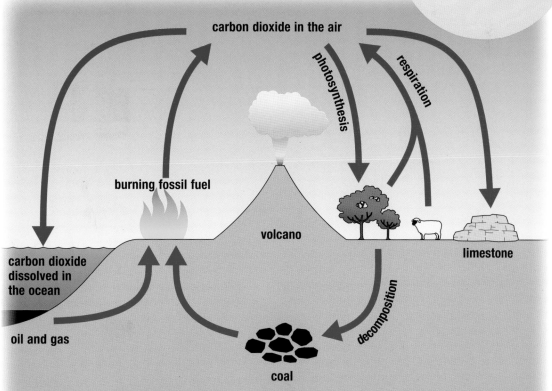

carbon dioxide in the air

photosynthesis

respiration

burning fossil fuel

volcano

carbon dioxide dissolved in the ocean

limestone

oil and gas

decomposition

coal

The movement of carbon and oxygen within the cycle takes place in different ways. Plants absorb carbon dioxide from the atmosphere every day and use it to make food and oxygen. Animals and humans eat the plants and so carbon gets into their bodies. In this way there is some carbon in all living things. When plants and animals break down their food to release **energy**, they take in oxygen and return carbon dioxide to the atmosphere.

When plants and animals die, their bodies are broken down by bacteria (tiny living things) and carbon is released into the soil. During this process, the bacteria use up oxygen and release carbon dioxide into the atmosphere.

Sometimes dead plants and animals may not **decay** fully and, over millions of years, they are changed into **fossil fuels** such as coal and oil. When fossil fuels are burned, carbon dioxide is released into the atmosphere faster than in the natural carbon-oxygen cycle. This upsets the balance of carbon dioxide and oxygen in the atmosphere. This may have important effects on the Earth's climate.

Carbon dioxide enters the atmosphere when volcanoes erupt.

The ocean absorbs carbon dioxide from the air and stores it. The
carbon dioxide may remain **dissolved** in the water, or it may be
used by tiny sea creatures to form their shells. When these animals
in the sea die, their bodies sink to the bottom and pile up into
sediments. Gradually these are squashed to form **sedimentary
rock** such as limestone. The carbon dioxide remains locked away
in these rocks for millions of years.

These limestone rocks
in Australia keep carbon
dioxide locked away for
millions of years.

What is the nitrogen cycle?

The cycling of nitrogen between plants, animals, and the atmosphere is called the **nitrogen cycle**. It is controlled by tiny bacteria in the soil, which change nitrogen from the atmosphere into a form that plants can use.

Plants need nitrogen to make proteins, which help them grow. Even though nitrogen gas makes up 78 per cent of the atmosphere, plants cannot use it in this form. Tiny bacteria, called nitrogen-fixing bacteria, live in the roots of some plants and absorb nitrogen gas from the atmosphere. They change it into a form that plants can use, called **ammonia**. This is called **nitrogen fixing**.

Plants get nitrogen from tiny bacteria in the soil.

Other bacteria, called nitrifying bacteria, also change nitrogen into another form that plants can use, called **nitrates**. The plants can take up ammonia and nitrates from the soil through their roots and use them to make protein.

In another part of the nitrogen cycle, nitrogen is returned to the atmosphere by denitrifying bacteria. These also live in the soil. The denitrifying bacteria change nitrates back into nitrogen gas. The gas is then released back into the atmosphere.

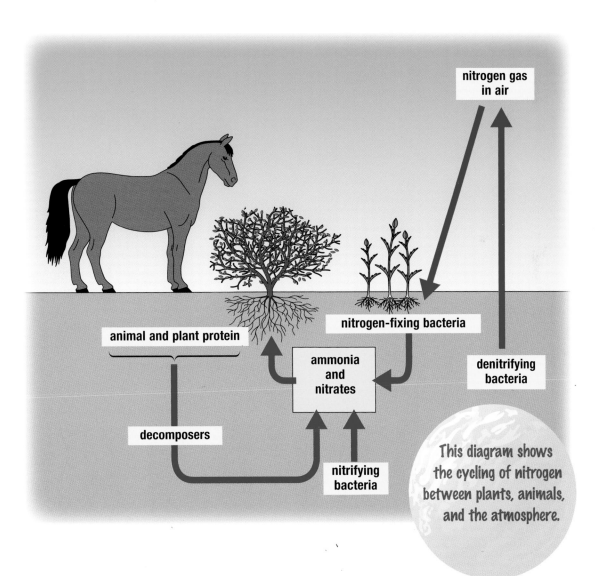

This diagram shows the cycling of nitrogen between plants, animals, and the atmosphere.

Animals and humans cannot get the nitrogen they need for movement and growth directly from the atmosphere or soil, so they get it by eating plants. When plants and animals die, they are broken down by bacteria called **decomposers**. The nitrogen stored in the plants and animals is recycled into the soil, so that the nitrogen cycle can begin again.

Lightning does the same job as nitrifying bacteria. It changes nitrogen in the atmosphere into nitrates which can be taken up by plant roots.

How does energy flow between living things?

As well as the cycling of carbon, oxygen, and nitrogen, the flow of energy between living things is also important for life on Earth. Energy is necessary for humans because it means we can move around and keep warm. We get energy from the food we eat.

The Sun provides energy for all living things through sunlight. The Sun's energy flows from one living thing to another in a **food chain**. Green plants are able to capture the Sun's light energy. They use it, together with carbon dioxide, to make food for themselves, and oxygen. Because they produce their own food, plants are called **producers**. They are the first stage in all food chains on Earth.

The Sun provides energy for all living things.

Animals cannot use the Sun's energy directly, so they get energy by eating plants and other animals. They are called **consumers**. Energy flows from the producer (the plant) to the consumer (the animal), but much of the energy is lost on the way.

When a plant or animal dies, decomposers help it decay. This returns energy to the environment so that it can be used again.

Did you know?

When an animal eats a plant, only 10 per cent of the energy eaten is stored in the animal's body. A massive 90 per cent is lost as heat and waste products.

This food chain shows the flow of energy on Earth.

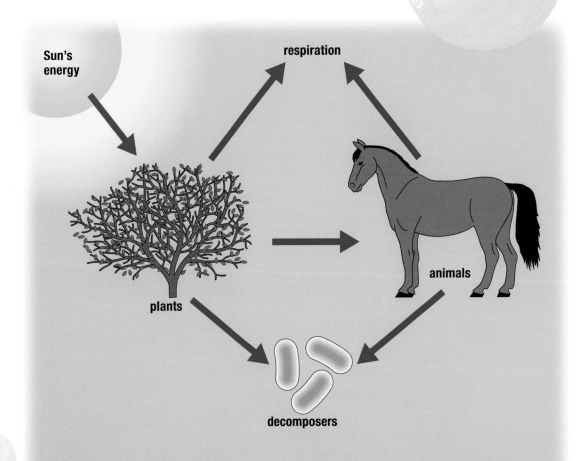

Sun's energy

respiration

plants

animals

decomposers

What is photosynthesis?

Plants are able to make their own food. They do this in a process called **photosynthesis**. It is one of the most important chemical reactions on Earth. It changes carbon dioxide into the food that we need to eat. It also produces the oxygen that we need to breathe.

Photosynthesis takes place in the leaves of green plants. Plants are green because they contain a chemical called **chlorophyll**, which can absorb light energy from the Sun. The plants use the Sun's energy, together with carbon dioxide from the air and water from the soil, to produce oxygen and food. Photosynthesis forms part of the carbon-oxygen cycle.

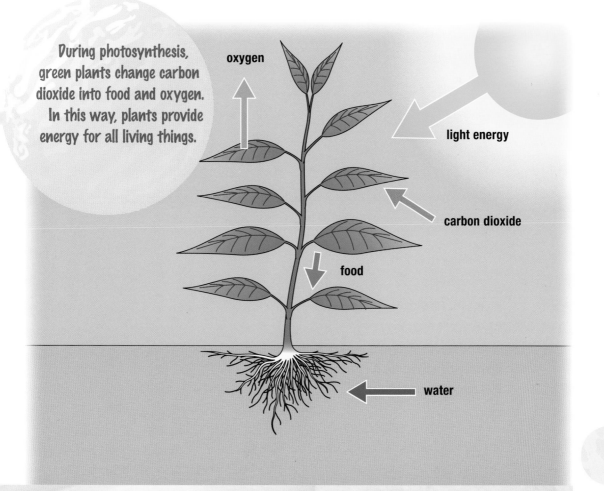

During photosynthesis, green plants change carbon dioxide into food and oxygen. In this way, plants provide energy for all living things.

oxygen

light energy

carbon dioxide

food

water

The food that plants make is stored inside them as energy called **carbohydrate**. When we eat fruit and vegetables, we take in some of this energy. If plants are not eaten, the energy stored inside them remains there when the plants die. Some plants that died in the days when the dinosaurs roamed the Earth, around 200 million years ago, gradually became buried and turned into fossil fuels such as coal and oil.

Did you know?

When we burn fossil fuels to heat our homes or create electricity, the energy that is made originally came from the Sun. Plants captured this energy in photosynthesis many years ago.

Coal is a fossil fuel. Millions of years ago these lumps of coal were plants, which photosynthesized just like plants do today.

When plants make food, they release oxygen into the air. Oxygen is necessary for life. If there was no oxygen we could not survive. Photosynthesis helps keep the balance of carbon dioxide and oxygen in the air. The huge areas of tropical rainforests in countries like Brazil are very important in helping to keep this balance.

Each year thousands of square kilometres of tropical rainforest are cut down and burned to clear land for farming. This is upsetting the balance. It leads to an increase in carbon dioxide in the air because fewer trees are available to take it in. Scientists think that this is forcing the temperature of the Earth to rise. It may have a harmful effect on the Earth's climate.

The tropical rainforests have been called "the lungs of the Earth" as they take in carbon dioxide and give off oxygen on a huge scale.

What is respiration?

When we breathe in oxygen from the atmosphere and when we eat fruit and vegetables we are gathering the ingredients for **respiration**. Respiration is when our bodies use oxygen to break down the food we eat and release the energy it contains. During respiration, the energy stored in the carbohydrates is released to produce carbon dioxide and water. We can then use this energy to walk, run, keep warm, and grow.

Respiration and photosynthesis are closely linked. Like photosynthesis, respiration plays a major role in keeping the balance of oxygen and carbon dioxide in the atmosphere. Respiration also forms part of the carbon-oxygen cycle.

Did you know?

Plants respire as well as photosynthesize.

Respiration occurs in all living things. It provides the energy needed for plants, animals, and humans to move and grow.

What is decomposition?

When plants and animals die, they decay. This means they become food for certain types of bacteria called **decomposers**. These are tiny organisms that eat dead plants and animals.

As the decomposers eat the dead plants and animals, they break them down into **nutrients**, such as nitrates, that are released back into the soil. This is called **decomposition**. Once in the soil, the nutrients dissolve in water and can then be taken up by plant roots and used again to make food. As the decomposers break down dead plants and animals, they also release carbon dioxide into the atmosphere during respiration. This can be used by plants in photosynthesis.

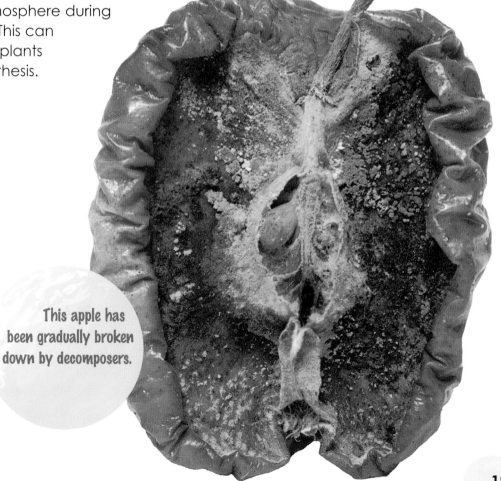

This apple has been gradually broken down by decomposers.

Did you know?

Nature continually recycles materials so that they can be used over and over again. All life is made up of recycled materials.

You will have noticed that in autumn the leaves fall from some types of trees, known as deciduous trees. The leaves pile up on the ground, but where do they go after that? They gradually disappear as they decompose and the nutrients are returned to the soil.

Death and decomposition play an important part in the cycling of nutrients such as carbon and nitrogen. Without decomposers, the world would be covered with dead plants and animals, and we would quickly run out of nutrients for new growth and life.

As decomposers get to work on these leaves the nutrients will be released back into the soil and the atmosphere.

How do humans affect the carbon-oxygen cycle?

Humans have affected the natural carbon-oxygen cycle for many years by burning fossil fuels and cutting down large areas of tropical rainforests. These activities release carbon dioxide back into the air much faster than would occur naturally.

Fossil fuels, such as coal and oil, are formed over millions of years from the dead remains of plants and animals. When these fossil fuels are burned, carbon dioxide is released into the atmosphere. As the population of the world increases, more and more fossil fuels are being burned. This means carbon dioxide is returned to the atmosphere more quickly than plants can remove it in photosynthesis.

When fossil fuels are burned in factories, carbon dioxide is released into the atmosphere.

Each year, thousands of square kilometres of tropical rainforests are cut down, particularly in Asia and South America, to make way for farming or buildings. Destroying forests means that there are fewer plants to absorb the increased amounts of carbon dioxide in the atmosphere.

Due to these kinds of human activities, the amount of carbon dioxide in the atmosphere has gradually increased over the last 200 years. It is still increasing.

Did you know?

In 1990 the level of carbon dioxide in the atmosphere was 350 parts per million (compared to 280 ppm in 1800). Carbon dioxide makes up only a tiny amount of the atmosphere, so scientists measure it in "parts per million". So 350 parts per million means 350 bits of carbon dioxide for every million bits of atmosphere. Scientists think that this level will double within this century.

The clearing of tropical rainforests means that there are fewer plants to absorb the increased amounts of carbon dioxide in the atmosphere.

Is the Earth warming up?

Even though carbon dioxide only makes up a tiny amount of the Earth's atmosphere, it has a very important effect on climate. Carbon dioxide is a **greenhouse gas**. This means that it traps some of the Earth's outgoing heat, like the glass roof of a greenhouse. As the amount of carbon dioxide in the air increases, it traps more heat, and so the Earth gets warmer. This is called the **greenhouse effect**.

This graph shows the level of carbon dioxide in the atmosphere since 1958. You can see that it has been gradually rising.

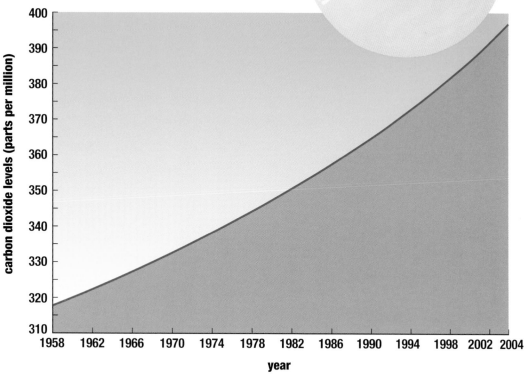

If the level of carbon dioxide in the atmosphere doubles during the 21st century, scientists think that the **temperature** of the Earth will increase by up to 5 °C. This rise in temperature is called **global warming**. People who live in cold countries may like the sound of this! But if the Earth warms up, it could lead to the melting of ice sheets, which are thick layers of ice that cover large areas of the North and South Poles. This could cause sea levels to rise by 1.5 metres (5 feet).

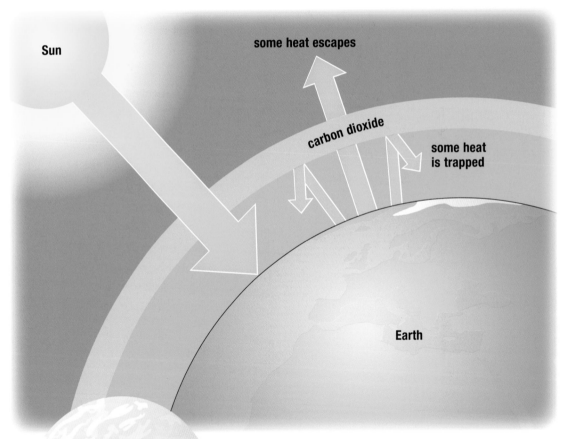

Sun

some heat escapes

carbon dioxide

some heat is trapped

Earth

Carbon dioxide in the atmosphere acts like the roof of a greenhouse as it traps some of the Earth's outgoing heat.

Did you know?

If sea levels rose by 1.5 metres (5 feet), this would lead to flooding in cities such as New York and London. Some low lying islands, such as the Maldives, would completely disappear under water.

Scientists are like detectives; they need to look at all the available information to work out what has happened to the Earth in the past. They have found out that the Earth has already warmed by 0.5 °C since 1900. You may not think this is very much, but it means that many **glaciers** in the mountains are beginning to melt. It also means that trees are now growing in northern Alaska, where it used to be too cold for anything to grow.

Scientists are not sure what will happen next because there are so many ways of removing carbon dioxide from the atmosphere. We do not know how much of the extra carbon dioxide produced by burning fossil fuels can be absorbed by the oceans and forests.

One thing is certain though: when carbon dioxide is released into the atmosphere faster than by natural processes, the carbon-oxygen cycle is disrupted. At some point in the future, this may cause problems for life on Earth.

Exit Glacier in Alaska, USA, is melting.

How do humans affect the nitrogen cycle?

In the natural nitrogen cycle there is a balance between the different forms of nitrogen in the air, in the soil, and in plants and animals. Human activities have changed this balance in two main ways.

When fossil fuels, such as coal and oil, are burned by humans, nitrogen is released into the atmosphere. This releases nitrogen back into the air much faster than would happen in the natural nitrogen cycle.

On farms, **fertilizers** containing nitrogen are added to crops to help them grow. As the world population increases, more fertilizers are used so that more food can be produced. This means extra nitrogen found in the fertilizer is added to the nitrogen cycle.

This plane is spraying the crop with fertilizer. This means that extra nitrogen enters the nitrogen cycle.

If too much fertilizer is used, it is washed off the fields when it rains, and into rivers and lakes. This is a waste of money for the farmer, and it also results in a process called **eutrophication**. This long word is the name given to the growth of bacteria and lots of tiny plants, called algae, in rivers or lakes. The algae and bacteria use up all the oxygen in the water, so fish and other animals living there cannot get enough oxygen and die.

This photograph shows algae growing in a river polluted with fertilizers.

Conclusion

Carbon, oxygen, and nitrogen are extremely important for life on Earth. They are recycled naturally between the soil, plants, animals, the air, and the ocean. Without this recycling, life on Earth would slowly but eventually stop.

The flow of energy between living things is important for life on Earth. The Sun provides energy for all living things. Green plants are able to capture the Sun's energy and use it to make food and oxygen through the process of photosynthesis. Animals cannot use the Sun's energy directly, so they get energy by eating plants and other animals.

When a plant or animal dies, it is broken down by decomposers. This process returns energy and nutrients to the environment so that they can be used again.

Humans have affected the natural cycles for many years by burning fossil fuels, cutting down large areas of tropical rainforests, and adding fertilizers to soils. The effects of these activities may lead to global warming.

The Sun's energy and the carbon, oxygen, and nitrogen in the atmosphere are important for life on Earth.

Fact file

Sink	Amount of carbon (billions of tonnes)
Atmosphere	766
Soil	1500
Ocean	40,000
Sedimentary rocks	100,000,000
Plants	600
Fossil fuels	4000

The level of carbon dioxide in the atmosphere.

Carbon dioxide level (parts per million)	Year
280	1800
315	1957
356	1993
365	2000
397	2004

Did you know?

On a hot day, a large tree takes in enough water through its roots to fill about five bathtubs. It has to take in this much to replace the water that it loses during respiration.

Glossary

absorb soak up, take in

ammonia form of nitrogen that plants can use

atmosphere blanket of air that surrounds the Earth

carbohydrate food that plants make

carbon-oxygen cycle movement of carbon and oxygen between the atmosphere, the oceans, life, and rocks

chlorophyll green chemical found in plants that can absorb sunlight

climate type of weather an area usually experiences

consumer animal that cannot use the Sun's energy directly, so gets its energy by eating plants and other animals

decay rot away

decomposer tiny organism that eats dead plants and animals

decomposition breaking down dead plants and animals

dissolve when a substance mixes with water and becomes part of it

element pure substance. Carbon, oxygen, and nitrogen are all elements.

energy power that is used to make something

eutrophication process where excess nitrogen enters rivers and lakes and causes lots of algae and bacteria to grow

fertilizer chemical added to crops on farms to help them grow

food chain description of the way that energy from the Sun flows from one living thing to another

fossil fuels fossilized remains of plants and animals that are millions of years old such as coal, oil, and natural gas

glacier slow-moving river of ice

global warming heating up of the Earth

greenhouse effect as the amount of carbon dioxide in the air increases, it traps more heat, and so the Earth gets warmer

greenhouse gas gas, such as carbon dioxide, that traps some of the Earth's outgoing heat (like the glass roof of a greenhouse)

lightning spark produced by a build-up of static electricity in a huge thunderstorm cloud

limestone sedimentary rock made from fossils on the ocean floor

nitrate form of nitrogen that plants can use

nitrogen cycle movement of nitrogen between plants, animals, and the atmosphere

nitrogen fixing absorbing nitrogen gas from the atmosphere and changing it into a form that plants can use (this is done by nitrogen-fixing bacteria)

nutrient food that plants and animals need

photosynthesis process where a plant uses the Sun's energy together with carbon dioxide from the air and water from the soil to produce oxygen and food

producer plant that can produce its own food at the beginning of a food chain

protein part of the food we eat that helps growth

recycle re-use

respiration process where plants and animals use oxygen to release the energy stored in the carbohydrates produced in photosynthesis (and also produce carbon dioxide and water)

sediment bits of rock and mud laid down

sedimentary rock rock formed from deposited broken bits of other rocks

sink place where carbon or oxygen is removed from the carbon-oxygen cycle and stored

temperature how hot or cold a place is

Index

Malpas

14.08.18

More books to read

Microhabitats: Life in a Garden, Clare Oliver (Evans, 2002)
Rotters, John Townsend (Raintree, 2005)
Science Answers: Food Chains and Webs, Richard and
Louise Spilsbury (Heinemann Library, 2004)